I love reading

To the Rescue

by Monica Hughes

Consultant: Mitch Cronick

BEARPORT
PUBLISHING COMPANY, INC.
New York, New York

Credits

t=top, b=bottom, c=center, l=left, r=right, OFC=outside front cover
Corbis: 16–17, 18. Oshkosh: 12, 13. Photodisc: 14. Superstock: 5, 6–7, 8, 10, 11, 19.
ticktock photography: 4, 7tr, 9. US Coastguard: 14–15, 20, 21.

Library of Congress Cataloging-in-Publication Data

Hughes, Monica.

    To the rescue / by Monica Hughes.

       p. cm. — (I love reading)

    Includes index.

    ISBN 1-59716-156-X (library binding) — ISBN 1-59716-182-9 (pbk.)

    1. Rescue work — Juvenile literature. I. Title. II. Series.

    TH9402.H84 2006

    628.9'2 — dc22

                       2005030629

For more information, write to Bearport Publishing Company, Inc., 101 Fifth Avenue, Suite 6R, New York, New York 10003.
Printed in the United States of America.

1 2 3 4 5 6 7 8 9 10

The I Love Reading series was originally developed by Tick Tock Media.

CONTENTS

Ambulances

An ambulance races to save people's lives.

Its red lights flash.

Its **siren** goes off.

Lights

WATER WITCH EMS
CECIL 793

AMBULANCE

4

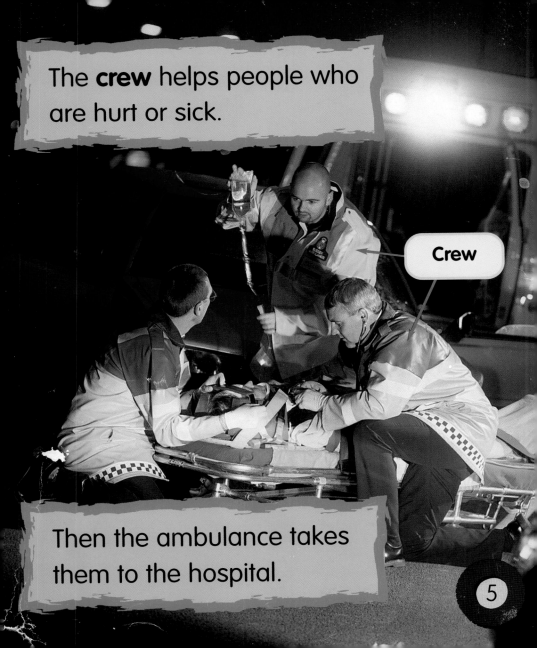

The **crew** helps people who are hurt or sick.

Crew

Then the ambulance takes them to the hospital.

5

Police

Police can help at an accident.

They have fast cars.

Their cars have a siren and lights that flash.

Police have fast motorcycles, too.

Accident

Lights

POLICE

W827 AKP

7

Fire engines

There is a big fire in the city.

A fire engine brings firefighters
to put out the fire.

Some fire engines have long ladders.

Ladder

Firefighters climb the ladders to rescue people in tall buildings.

9

Firefighters

Fire engines have long **hoses**.

The hoses can spray water.

Ladder

Hose

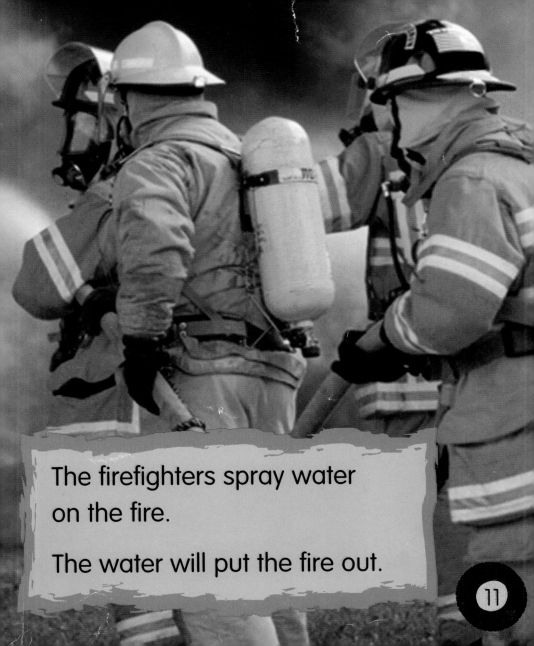

The firefighters spray water on the fire.

The water will put the fire out.

Airport fires

There is a fire at the airport.

The Striker goes to help.

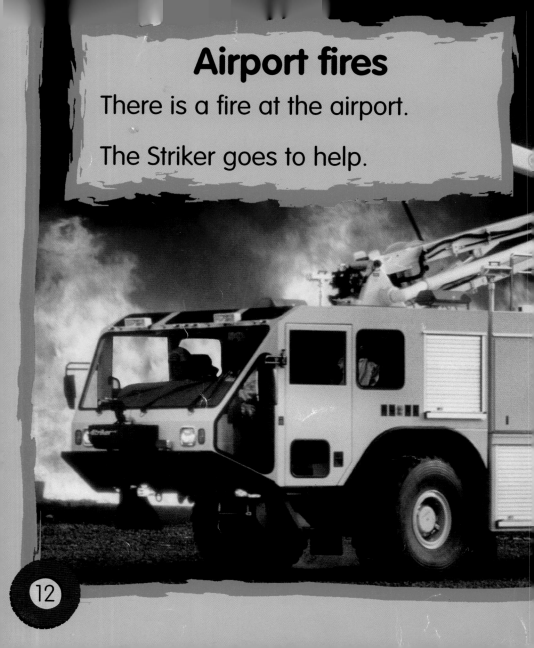

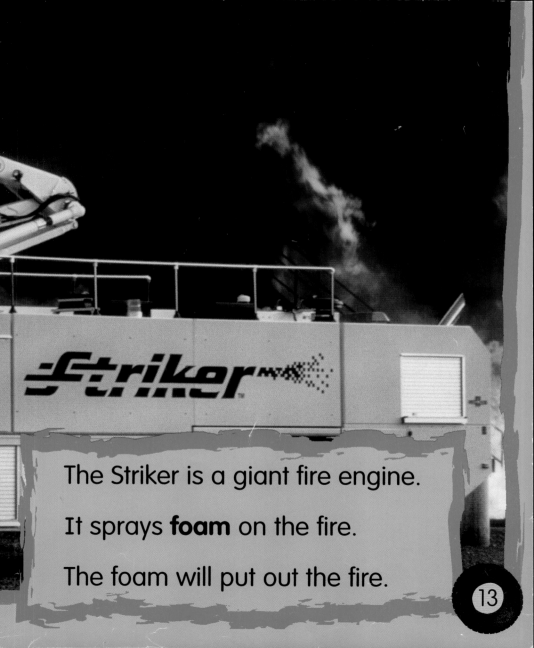

The Striker is a giant fire engine.

It sprays **foam** on the fire.

The foam will put out the fire.

13

Forest fires

There is a fire in a forest.

An air tanker flies over the forest.

Forest fire

It drops powder on the fire.

The powder stops the forest fire from getting bigger.

Fireboats

A fireboat puts out fires at sea and on rivers.

It sprays water on the fire.

The fireboat gets the water it uses from the sea or the river.

Snowplows

This road is blocked by snow.

Cars cannot get by.

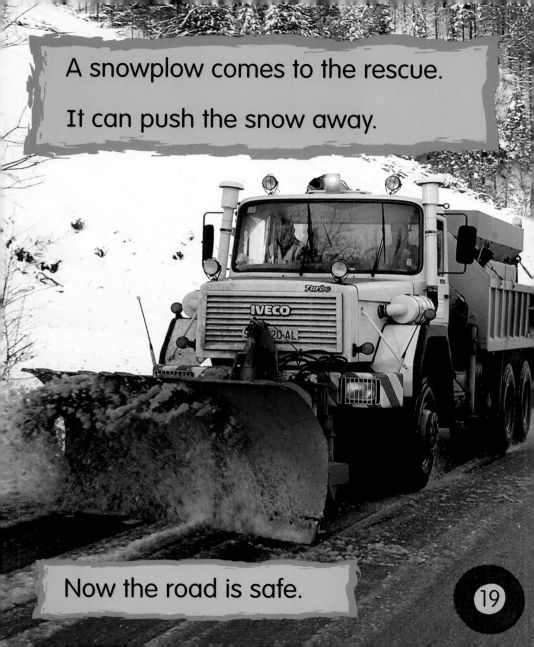

A snowplow comes to the rescue.

It can push the snow away.

Now the road is safe.

19

Helicopters

A helicopter can help rescue
people lost at sea.

It can **hover** in the air.

Then the crew can drop down a rope to save someone in trouble.

Glossary

crew (KROO)
a group of people
who work together
to get a job done

foam (FOHM)
lots of very
small bubbles
made from
chemicals

hoses (HOHZ-ez)
long rubber tubes

hover (HUHV-ur)
to stay in one place
in the air

siren (SYE-ruhn)
a machine that
makes a loud sound
to warn people

23

Index

Learn More

Kottke, Jan. *A Day with Firefighters.* Danbury, CT: Children's Press (2000).

Liebman, Dan. *I Want to Be a Police Officer.* Ontario, Canada: Firefly Books Ltd. (2000).

www.hunterlimo.com/web_safe_kids/safe_kids.asp

www.nfpa.org/sparky/firetruck/gallery/gallery.htm

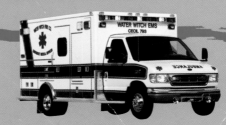